A **TRUE** BOOK™

UNDERSTANDING CLIMATE CHANGE

The Greenhouse Effect

Mara Grunbaum

Children's Press®
An Imprint of Scholastic Inc.

Content Consultant
Heidi A. Roop, PhD
Research Scientist
Climate Impacts Group
University of Washington, Seattle
Seattle, Washington

Library of Congress Cataloging-in-Publication Data
Names: Grunbaum, Mara, author.
Title: The greenhouse effect/Mara Grunbaum.
Other titles: True book.
Description: New York: Children's Press, an imprint of Scholastic Inc. [2020] | Series: A true book |
    Includes index. | Audience: Grades 4–6. (provided by Children's Press.)
Identifiers: LCCN 2019031393 | ISBN 9780531130766 (library binding) | ISBN 9780531133767 (paperback)
Subjects: LCSH: Greenhouse effect, Atmospheric—Juvenile literature. | Global warming—Juvenile
    literature. | Climatic changes—Juvenile literature. | Carbon dioxide—Juvenile literature.
Classification: LCC QC912.3 .G785 2020 | DDC 363.738/742—dc23 LC

Design by THREE DOGS DESIGN LLC
Produced by Spooky Cheetah Press
Editorial development by Mara Grunbaum

All rights reserved. Published in 2020 by Children's Press, an imprint of Scholastic Inc.
Printed in North Mankato, MN, USA   113

Scholastic Inc., 557 Broadway, New York, NY 10012

1 2 3 4 5 6 7 8 9 10 R 29 28 27 26 25 24 23 22 21 20

**Front cover: Heavy traffic contributes
to the greenhouse effect.**

Back cover: A scientist inspects an ice drill.

Find the Truth!

Everything you are about to read is true *except* for one of the sentences on this page.

Which one is **TRUE**?

T or F Carbon dioxide in the atmosphere lowers Earth's temperature.

T or F Scientists have observed Earth getting warmer in recent decades.

Find the answers in this book.

Contents

Energy from the sun gives Earth heat and light.

The BIG Truth

Your Carbon Footprint

4 Heating the Planet

A climate-monitoring station

A Critical Moment

In recent centuries, humans have released increasing amounts of **greenhouse gases** into Earth's atmosphere. These gases have trapped heat in the atmosphere, causing the average temperature on the planet's surface to rise and contributing to **global warming**.

Because of global warming, oceans are heating up, sea levels are rising, and weather is becoming more extreme. These changes in Earth's climate are known as global **climate change**. They threaten people and other plant and animal species around the world. If we don't make changes to reduce greenhouse gas **emissions** now, these problems will worsen.

There is good news, though!

Thousands of scientists worldwide are studying global climate change. Politicians, public figures, and citizens of all ages are trying to figure out what to do. Humanity now knows more than ever about the causes and effects of climate change, as well as how we might reduce its impact. That means **people today can make decisions** that will affect the planet for centuries to come.

Turn the page to learn how greenhouse gases in Earth's atmosphere affect the climate through a phenomenon called the greenhouse effect.

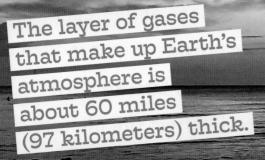

Sunlight being scattered by particles in the atmosphere creates the colors of a sunset.

The layer of gases that make up Earth's atmosphere is about 60 miles (97 kilometers) thick.

1

Our Atmosphere

Next time you go outside, look up. You might see blue sky, gray clouds, brown haze, or a stunning red sunset. All of this is taking place in Earth's atmosphere. This layer of invisible gases, water vapor, and dust particles surrounds the planet we live on. Billions of humans, animals, and plants depend on the atmosphere. It provides the air we breathe, and without it, humans would not be able to live on Earth.

The atmosphere keeps Earth at an average temperature of 59 degrees Fahrenheit (15 degrees Celsius).

Sun Shade

The sun is constantly pumping out energy. This energy travels 93 million miles (150 million km) to our planet in the form of light. The atmosphere filters out most of the harmful parts of the sun's rays. The light that does make it through provides energy for every living thing on Earth. Plants use energy from sunlight to make their own food through **photosynthesis**. People and animals eat plants to gain energy for themselves.

Invisible Blanket

Light from the sun also bathes Earth's surface with warmth. During the day, sunlight heats up the water and land. At night, Earth's surface cools again, releasing heat upward. But the atmosphere acts like a blanket, trapping most of this heat before it escapes into space. This phenomenon is called the greenhouse effect.

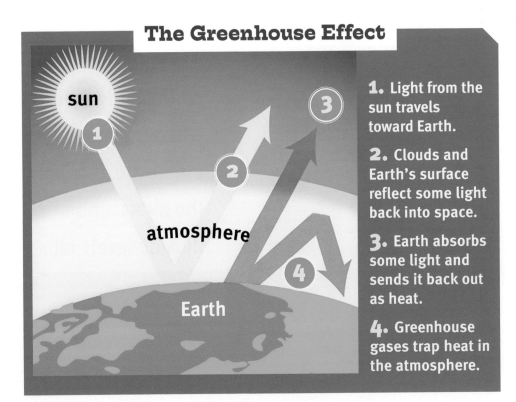

The Greenhouse Effect

sun

atmosphere

Earth

1. Light from the sun travels toward Earth.

2. Clouds and Earth's surface reflect some light back into space.

3. Earth absorbs some light and sends it back out as heat.

4. Greenhouse gases trap heat in the atmosphere.

What's in the Atmosphere?

The atmosphere is composed mostly of nitrogen and oxygen. These two gases make up about 99 percent of the air. There are smaller amounts of many other gases, including argon, methane, **carbon dioxide** (CO_2), and water vapor. This mix naturally changes as Earth's surface and atmosphere interact. For example, plants take in carbon dioxide from the air as part of photosynthesis. In spring and summer, when the most plants are growing, the air's carbon dioxide levels fall.

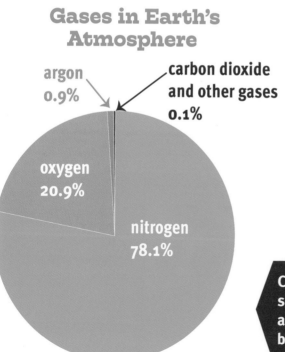

Gases in Earth's Atmosphere

argon 0.9%

carbon dioxide and other gases 0.1%

oxygen 20.9%

nitrogen 78.1%

Carbon dioxide is a small part of Earth's atmosphere, but it has a big impact on the planet.

The greenhouse effect is named after greenhouses, which trap heat to help plants grow.

Greenhouse Gases

Gases such as carbon dioxide, methane, and water vapor are greenhouse gases. Think about the "atmosphere as a blanket" analogy. When there are fewer greenhouse gases in the atmosphere, the blanket is thinner, and there is less warming from the greenhouse effect. But as we add greenhouse gases to the atmosphere, the blanket becomes thicker. More heat gets trapped in Earth's atmosphere.

A warmer atmosphere can make the land hotter and drier in many places.

Climate Control

Greenhouse gases regulate the temperature of Earth's atmosphere. The heat they trap can be absorbed by the land and oceans, warming those, too. In recent decades, humans have released more greenhouse gases into Earth's atmosphere, mostly by burning fuel for cars, trucks, and airplanes. As a result, average temperatures around the globe are going up. This global climate change affects weather, sea levels, and living things everywhere.

Alien Atmosphere

Earth is the only planet we know of that can support life. That's largely because our atmosphere keeps Earth at the right temperature for life to survive. Venus is our closest neighbor. Its atmosphere has 154,000 times more carbon dioxide than Earth's does. Venus is only slightly closer to the sun than Earth is. But the greenhouse effect heats its surface to 880°F (471°C)—hot enough to melt lead!

The hot surface of Venus is shown in this artist's illustration.

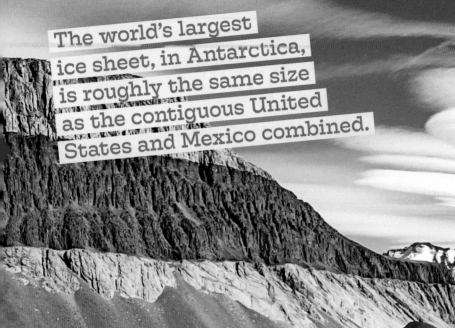

The world's largest ice sheet, in Antarctica, is roughly the same size as the contiguous United States and Mexico combined.

A scientist checks drilling equipment in Antarctica.

A History of the Climate

Hundreds of thousands of years ago, no one was measuring the gases in Earth's atmosphere. But scientists have ways of studying what the planet was like back then. For example, researchers drill into frozen **ice sheets** to collect samples called ice cores. These ice sheets formed as fallen snow was packed down every year. The process trapped bubbles of air from the atmosphere. Scientists analyze these bubbles to see how the atmosphere has changed over time.

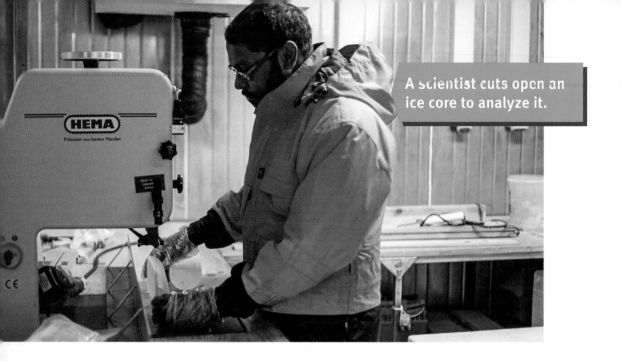

A scientist cuts open an ice core to analyze it.

Climate Clues

Ice cores hold additional clues about Earth's climate in the past. Snow and ice form differently depending on how cold it is. Analyzing the layers can tell scientists how the temperature changed from year to year. The thickness of each layer also shows how much snow fell. Trapped ash and soot tell scientists that there were volcanic eruptions or forest fires. This ice core evidence helps scientists reconstruct Earth's climate history.

Natural Cycles

Ice cores show that Earth's temperature has changed before. One reason is that Earth's **orbit** around the sun shifts slightly over time. This changes the amount of sunlight Earth receives, making the planet warmer or cooler. Over thousands of years, these shifts have warmed and cooled Earth repeatedly. During cooler periods, called ice ages, ice covered large portions of the planet. During warmer periods, the climate conditions were similar to those today.

Woolly mammoths roamed Earth during the last ice age.

Earth's last ice age ended about 12,000 years ago.

Extreme Eruptions

Other natural changes have happened suddenly. In 1815, for example, a volcano in Indonesia spewed gas and ash into Earth's atmosphere, temporarily blocking out sunlight. The Northern Hemisphere's average temperature dropped by about 3°F (1.7°C) for the next year. In parts of the United States, snow fell in July. But the temperature returned to normal once the gas and ash cleared.

Erupting volcanoes release plumes of gas and ash.

Unusual Events

Now Earth's climate is changing again. Since the late 1800s, the average temperature on Earth's surface has risen 1.8°F (1°C). This change is too fast to be caused by shifts in Earth's orbit. It's too long-lasting to be explained only by natural forces such as volcanoes. Instead, scientists point to rising levels of greenhouse gases in Earth's atmosphere. Our atmosphere contains more of these gases today than at any time in at least 800,000 years.

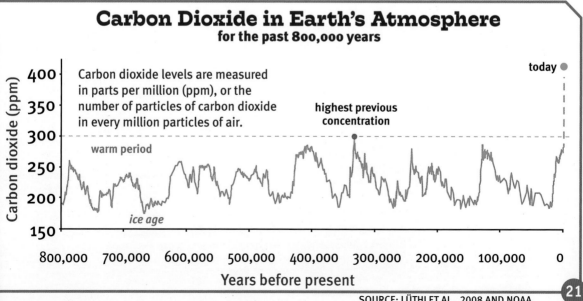

Carbon Dioxide in Earth's Atmosphere
for the past 800,000 years

Carbon dioxide levels are measured in parts per million (ppm), or the number of particles of carbon dioxide in every million particles of air.

today

highest previous concentration

warm period

ice age

Carbon dioxide (ppm): 400, 350, 300, 250, 200, 150

Years before present: 800,000, 700,000, 600,000, 500,000, 400,000, 300,000, 200,000, 100,000, 0

More than two billion people around the world use energy from burning wood to heat their homes and to cook.

Burning wood releases energy in the form of light and heat.

Energy from Earth

Humans have lived on Earth for about 300,000 years. Over time, people have found many ways to use energy from nature. Since prehistoric times, humans have burned wood and animal dung to cook and keep warm. Early people used the sun's energy to dry grain and other goods. At least 5,000 years ago, people tamed horses for riding and transporting goods. Wind was used to power sailboats. Running water from streams and rivers powered early machines.

The coal people use today took millions of years to form.

Ancient Fuel

More than 4,000 years ago, people in China discovered a new energy source. They collected chunks of coal found near Earth's surface. Burning it released heat and light. Coal was formed when plants and animals died in ancient swamps. Their remains were gradually covered in layers of dirt and sand. Over millions of years, heat and pressure beneath Earth's surface compressed them into hard rock. Similar processes formed other **fossil fuels**, including oil and natural gas.

The Industrial Revolution

Most fossil fuels are buried deep beneath Earth's surface. In the late 1500s, people in England started mining coal from underground. Within two centuries, as mining techniques improved, coal became the main fuel source throughout Europe. It powered the first factories, which produced tools and clothing formerly made in people's homes. This was the beginning of a period called the **Industrial** Revolution. People in Europe and the United States started using more fossil fuels.

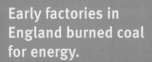
Early factories in England burned coal for energy.

Modern Times

As the Industrial Revolution continued, coal-powered ships and trains were invented. People learned to drill for buried oil and natural gas. These energy sources fueled many new technologies. Today, most human activities are powered by fossil fuels. Most cars run on gasoline, which is made from oil. More than half of the world's power plants burn coal and natural gas. The electricity they make heats homes and runs factories. It powers everything from refrigerators to phones.

Timeline: Use of Fossil Fuels

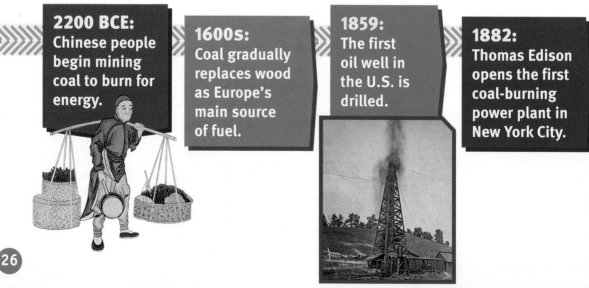

2200 BCE: Chinese people begin mining coal to burn for energy.

1600s: Coal gradually replaces wood as Europe's main source of fuel.

1859: The first oil well in the U.S. is drilled.

1882: Thomas Edison opens the first coal-burning power plant in New York City.

Fueling Pollution

Using fossil fuels for energy has downsides. Coal, oil, and natural gas are not **renewable**. They took millions of years to form naturally. If people keep using them at current rates, these fuels could run out in 50 to 100 years. And there's another problem that scientists say is even more urgent: Burning fossil fuels releases carbon dioxide and other greenhouse gases into the atmosphere. This strengthens the greenhouse effect, which leads to global climate change.

1885:
The first gasoline-powered car is built in Germany.

1913:
The Ford Motor Company begins mass-producing cars in Detroit, Michigan.

1939:
The first jet aircraft, powered by diesel fuel, takes flight.

TODAY:
Fossil fuels power most vehicles and more than half of the world's power plants.

Your Carbon Footprint

The amount of carbon dioxide that comes from your own day-to-day activities is called your carbon footprint. Here's how your choices can affect your carbon footprint, according to scientists who have done the math.

Getting Around

A 10-mile (16.1 km) drive in an average car releases 8.9 pounds (4 kilograms) of carbon dioxide. Walking, biking, or taking public transit instead of driving can help reduce this carbon dioxide in the atmosphere.

Preparing Meals

Growing plants takes less energy than raising animals like cows, pigs, sheep, and chickens. If you eat vegetables instead of meat once a week, you could save about 660 pounds (300 kg) of carbon dioxide in a year.

Buying Things

Manufacturing takes a lot of energy. Making just 1 pound (0.5 kg) of plastic releases more than 1 pound (0.5 kg) of carbon dioxide. Sharing toys and other items with friends or buying used products instead of new can help reduce this impact.

About 80 percent of the energy used in the United States comes from fossil fuels.

Offshore oil platforms use long drills to reach oil deposits beneath the ocean floor.

Heating the Planet

People now use over a thousand times more fossil fuels than they did before the Industrial Revolution. Demand for energy has risen as more countries have become industrialized and the world's population has grown. In 2018, burning fossil fuels released 36.5 billion tons (33.1 billion metric tons) of carbon dioxide. Some of it was absorbed by Earth's oceans and land surface. But the rest stayed in the atmosphere, where it can remain for thousands of years.

A Growing Problem

In 1958, American scientists started monitoring the carbon dioxide in Earth's atmosphere from the Mauna Loa Observatory, high up on a mountain in Hawaii. Every day since then, instruments at the observatory have collected air samples and analyzed what is in them. After gathering these data for more than 60 years, scientists see a clear pattern. The average amount of carbon dioxide in the atmosphere has been steadily going up.

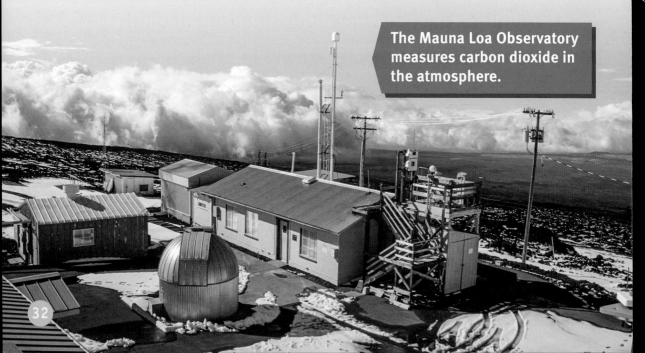

The Mauna Loa Observatory measures carbon dioxide in the atmosphere.

Measurements from research stations and ocean buoys around the world also show that global average temperature is rising. The atmosphere is warming about ten times faster than it did at the end of the last ice age. Organizations around the world have reported that this is happening. A majority of scientists and government agencies agree that this warming can't be explained

A typical car releases 5.1 tons (4.6 metric tons) of carbon dioxide per year.

by natural processes alone. It is caused by the increased amount of greenhouse gases released by human activities.

Cutting down trees contributes to climate change.

Changing the Land

About three-quarters of humanity's greenhouse gas emissions come from fossil fuels. But people have changed the atmosphere in other ways, too. People have cut down nearly half of Earth's forests in order to build farms and cities. When there are fewer trees to take in carbon dioxide, more stays in the atmosphere. The Amazon rain forest in Brazil stores a lot of greenhouse gases. But one-sixth of the Amazon has been cut down in the past 50 years.

Excuse You!

About 15 percent of global greenhouse gas emissions come from livestock. Cows are the main culprits because of how they digest food. A cow's stomach contains microbes that help it break down hay and grasses. But this process creates methane, which the cows release when they burp. Methane can trap even more heat than other greenhouse gases. Some experts are studying whether feeding cattle differently can reduce this unpleasant effect.

A special backpack measures methane from a cow's stomach.

Future Predictions

Scientists know Earth is getting warmer. But many factors affect how warm it might get. To understand what could happen, scientists use computers that **simulate** Earth's climate. They show that global average temperatures could rise between 2.5°F (1.4°C) and 10°F (5.6°C) over the next century. Some places could get even hotter. What happens largely depends on how much more carbon dioxide we add to the atmosphere. The more Earth warms, the worse the effects will be on humans and other living things.

Scientists track weather patterns to understand the climate.

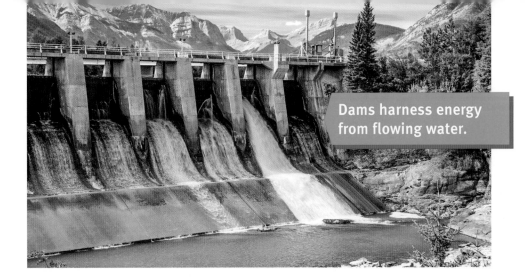

Dams harness energy from flowing water.

Energy Options

The most effective way to limit climate change is to reduce the use of fossil fuels. Some power companies now harness energy from sources that don't release carbon dioxide. Sunlight, wind, flowing water, and heat beneath Earth's surface provide renewable energy. Unlike fossil fuels, these natural processes will not run out. Nuclear power uses energy stored in atoms to make electricity without releasing greenhouse gases. But it requires uranium mined from Earth and is not considered renewable.

Making the Switch

Alternatives to fossil fuels are becoming more common. Nuclear power and renewable sources provided one-fifth of U.S. energy in 2018. But change has been slow, partly because most power plants and vehicles were designed to run on fossil fuels. Energy companies also make a lot of money extracting and selling them. Nuclear energy is popular worldwide, but accidents at nuclear plants can be very dangerous. That makes some people concerned about building more of them.

Workers install solar panels on the roof of a building.

Students in South Korea attend a march to urge action on climate change.

Now What?

Not everyone agrees on what to do about global climate change. Many people believe that the threat is urgent and that governments should act to reduce greenhouse gas emissions immediately. Others don't think that people should have to do things any differently. The future will depend on what today's societies decide. Scientists know that Earth's climate has changed before. But this time, humans are causing it—and humans have the power to affect what happens next. 🌍

CO_2 in the Atmosphere

Scientists collect all sorts of information related to our planet. They might measure sunlight, air or water temperature, or the makeup of the atmosphere. They look for patterns in these data to understand how the climate is changing. They share the information so people can study it for themselves. The graph at right shows carbon dioxide measurements taken at Mauna Loa. Study the graph, then answer the questions.

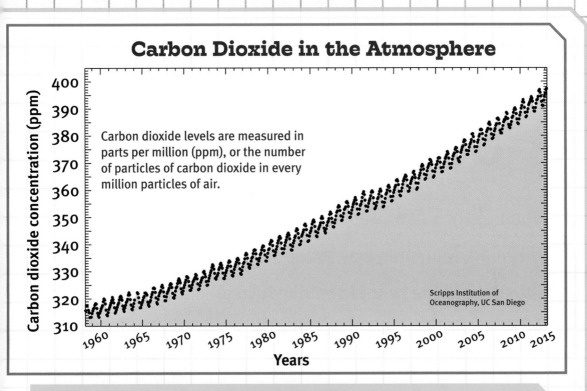

Carbon Dioxide in the Atmosphere

Carbon dioxide levels are measured in parts per million (ppm), or the number of particles of carbon dioxide in every million particles of air.

Scripps Institution of Oceanography, UC San Diego

Carbon dioxide concentration (ppm)

Years

Analyze It!

1 What was the concentration of carbon dioxide in the atmosphere in 1958?

2 What was the concentration of carbon dioxide in the atmosphere in 2015?

3 What pattern(s) do you see in the graph?

4 Think about what you've learned about the greenhouse effect. Based on this graph, would you predict that Earth's average temperature was warmer or cooler in 2019 than in 1958?

Mauna Loa Observatory, Hawaii

Solar Schools

When Claire Vlases was in seventh grade, her school in Bozeman, Montana, was scheduled to be renovated. She thought the school could reduce greenhouse gas emissions by installing solar panels on the roof.

Solar panels collect sunlight and convert it into electricity.

MAKING CHANGE
Claire Vlases worked with her school principal to have solar panels installed at her Montana middle school.

Claire asked the principal, who said it might be too complicated and expensive. But Claire decided to make the case. She gave a presentation about solar energy to the school board. She recruited other students, who raised $11,000 for the cause. In 2018, the school installed enough solar panels to provide one-quarter of its electricity. It's now one of nearly 5,500 U.S. schools using solar power.

Young people around the nation have pushed for changes in how their communities use energy. You can, too! Talk to school officials, write to your government representatives, or volunteer with a local organization. Small steps can add up to big changes.

True Statistics

Earth's average temperature rise since the late 19th century: 1.8°F (1°C)

The hottest year since people started keeping records (as of 2019): 2016

Maximum concentration of carbon dioxide in Earth's atmosphere for 800,000 years, according to ice core measurements: 300 ppm

Level of carbon dioxide in Earth's atmosphere in May 2019, measured at Mauna Loa: 415 ppm

Percentage of U.S. carbon dioxide emissions that come from transportation: 29

Amount of carbon dioxide people added to the atmosphere in 2018: 36.5 billion tons (33.1 billion metric tons), the most ever

Percentage of climate scientists who agree that the climate is changing because of human activities: 97

Did you find the truth?

(F) Carbon dioxide in the atmosphere lowers Earth's temperature.

(T) Scientists have observed Earth getting warmer in recent decades.

Resources

Other books in this series:

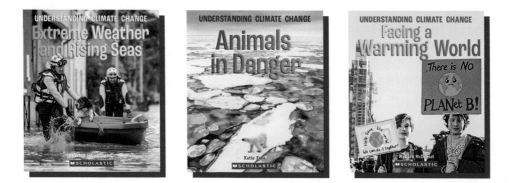

You can also look at:

Graham, Ian. *The Science of Weather: The Changing Truth About Earth's Climate*. New York: Children's Press, 2018.

Hamilton, Robert. *What's Climate Change?* New York: KidHaven Publishing, 2018.

Hirsch, Rebecca. *Climate Change and Energy Technology*. Minneapolis: Lerner Publications, 2019.

Scibilia, Jade Zora. *Climate Change* (Spotlight on Weather and Natural Disasters). New York: PowerKids Press, 2019.

Glossary

carbon dioxide (KAHR-bun dye-AHK-side) a gas that is a mixture of carbon and oxygen, with no color or odor

climate change (KLYE-mit chaynj) global warming and other changes in the weather and weather patterns that are happening because of human activity

emissions (i-MISH-uhnz) substances released into the atmosphere

fossil fuels (FAH-suhl FYOO-uhlz) coal, oil, and natural gas, formed from the remains of prehistoric plants and animals

global warming (GLOW-buhl WAR-ming) rise in temperature around Earth due to heat from the sun trapped by greenhouse gases in the atmosphere

greenhouse gases (GREEN-hous GAS-ez) gases such as carbon dioxide and methane that contribute to the greenhouse effect

ice sheets (ise sheets) thick sheets of ice that cover large areas of Earth for long periods of time

industrial (in-DUHS-tree-uhl) of or having to do with factories and making things in large quantities

orbit (OR-bit) the curved path followed by a moon, planet, or satellite as it circles a planet or the sun

photosynthesis (foh-toh-SIN-thi-sis) a chemical process by which green plants and some other organisms make their food

renewable (ri-NOO-uh-buhl) able to be replaced naturally over a short period of time

simulate (SIM-yuh-layt) to imitate real conditions or events

Index

Page numbers in **bold** indicate illustrations.

About the Author

Mara Grunbaum is an award-winning science writer for children and adults. She's the former editor of *SuperScience*, Scholastic's science magazine for students in elementary school. She has written about everything from giant sinkholes to the physics of roller coasters, but she's most fascinated by planet Earth and the variety of living things that populate it. She lives in Seattle, Washington, where she loves going for walks at low tide and playing with Zadie, the world's smartest cat.